THE SECRET LIFE OF
SCREECH OWLS!
NATURES TINY TERROR!
BY STEVE STONE

The Secret Life Of Screech Owls

By Steve Stone

Dedication:

This book was written to help people understand the unique form of life we call the 'screech owl,' and also to help raise funds for The American Wildlife Refuge, a raptor rescue center.

I wish to thank the volunteers and friends of The American Wildlife Refuge.
A special thanks to the board members of the AWR: Dr. Dan Johnson, the veterinarian who has worked with us for almost 20 years; Dan Notch, who brought falconry knowledge and 30+ years of experience; Chris & Cindy Bowman, who helped us connect with the cultural significance of our work; and Kelley Ferrante, who keeps us grounded in the real world, and also keeps reminding us of the wonder of the animals we help.
Finally, I wish to especially thank my wife Anitra, who has been my support both emotionally and financially while I spend year after year of my life helping animals and making no income because of it.

The Secret Life of Screech Owls

Published by Virtual Reflections Publishing
4501 New Bern Avenue, Suite 130-333
Raleigh, NC 27610

For more copies of this book, or to arrange author visits, please email Steve Stone at: AWRefuge@aol.com
Author Web Site: www.SteveStone.net
American Wildlife Refuge website: www.awrefuge.org
Publisher website: www.vreflections.com

The Secret Life Of Screech Owls

By Steve Stone

CONTENTS:

The Secret Life Of Screech Owls

By Steve Stone

1- What is a Screech Owl

A screech owl is any one of 29 of the smaller species of owl that are of the Megascops genus (See Appendix A). Screech owls average between 4 and 6 inches from perched feet to head, or 6 to 10 inches from stretched out head to tail. They are an eared owl, as they have feather tufts at the top of their head giving the impression of small pointed ears. Screech owls all have yellow to yellow green eyes. These tiny owls are spread out all throughout the Americas.

Just a few years ago, Eastern screech owls were called 'Otus' Asio, and Western were 'Otus kennicotti,' and so on with all screech owl species. Many older books still list the 'Otus" name rather than the newer Megascops name. The 'Otus' name was replaced with the 'Megascops' name to match screech owls with the similar 'Scops' owl species in Europe Asia, and Africa. Of the 29 Megascops species that are found in the Western hemisphere, only 3 are found in The United States. The 3 American screech owls are: the Eastern screech owl (Eastern and central US), the Western screech owl (Western US), and the whiskered screech owl (Southeast Arizona).

In The Eastern US (where this book is written) we have the Eastern screech owls. The Eastern screech owl's current scientific species name is Megascops Asio. The Eastern screech owls have 2 primary color variants. There are red phase screech owls, who appear bright reddish orange, and there are gray phase screech owls, who appear primarily gray and white. Both of the color variants change their feather colors as they age. Unlike humans whose hair turns gray with age, screech owls' feathers turn brown. A gray phase screech owl will become more and more brownish as it ages, until around 10 or more years old when it is no longer gray at all, but all brown. The same applies to the bright red screech owls. Their color seems to dim as the reds soften to red-brown, and finally just brown.

Eastern Screech Owls

In the western US we have the Western screech owls. The western screech owl's scientific species name is Megascops kennicotti. The Western screech owls are all almost exclusively gray. These gray birds remain the same gray their entire life. There is a subspecies of the Western screech owl in the San Fernando Valley where there are rufous (Brown) colored western screech owls, but they are rarely seen and low in numbers.

Western Gray Screech Owl

In southern Arizona down to Mexico, we have the whiskered Screech owl. The Whiskered screech owl's scientific species name is Megascops trichopsis. They also have a gray and a reddish brown phase, but they are slightly smaller than the Eastern and the Western screech owls. These owls are mostly in Mexico, though some are found in the US.

Although they are all called "screech owls," none of them specifically screech. The Eastern Screech owl makes a horse whinny sound and a musical trilling. The Western screech makes a bouncing ball like sound, or a musical trill similar to the eastern screech owl. The whiskered screech makes a "bububBUB" sound, which are a series of hoot like noises, the 3rd being the most emphasized.

In general all screech owls will look similar to most untrained eyes and through binoculars. The main thing to remember is they are less than 8 inches tall, they have yellow eyes, and they have feather 'ears' when they are not in full molt.

Red Eastern screech owl in full molt

2- The Babies

Like all birds, screech owls are born from eggs. Screech owl eggs are bright white slightly oval, but have a thinner end like chicken and many other eggs. The eggs are very small, but in comparison to the birds that lay them, they are huge! The 1 inch diameter, inch and a half long eggs come from an animal less than 6 inches tall!

From these come very tiny owls. They are so small that you can hold 4 of them in the palm of your hand.

Tiny babies just 2 inches tall!

These young owls will have voracious appetites. They have less than 30 days to grow from less than an inch to nearly 8 inches! They will eat and struggle for food to get themselves full size as quickly as possible.

Baby screech owl eating from forceps

Often there are 2 to 5 babies in the yearly nest.

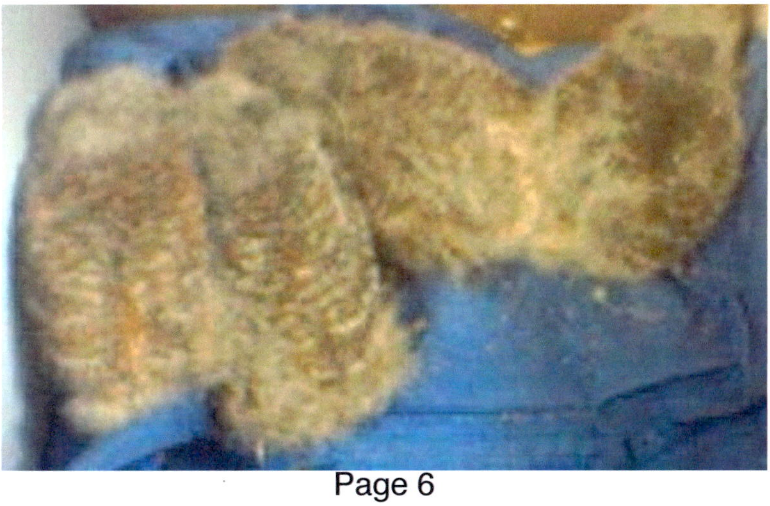

4 baby screech owls in an incubator

As they grow, their eyes open and their claws become more usable. They grip food (and fingers) and they watch everything!

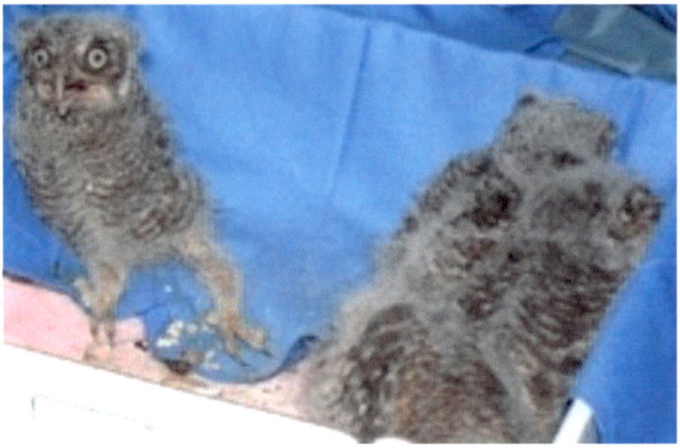

Same 4 screech owls, several days later.

Their feathers change from pure fluff to colored stripes and they begin to get their flight feathers. The flight feathers start off growing inside tiny little sheathes, making them look like small straws growing out of the bird.

Flight feathers beginning to grow.

They grow quickly, forming into small versions of the adults within a few weeks. The feathers now in place slowly push out the initial downy fluff, and the ear tufts begin to form.

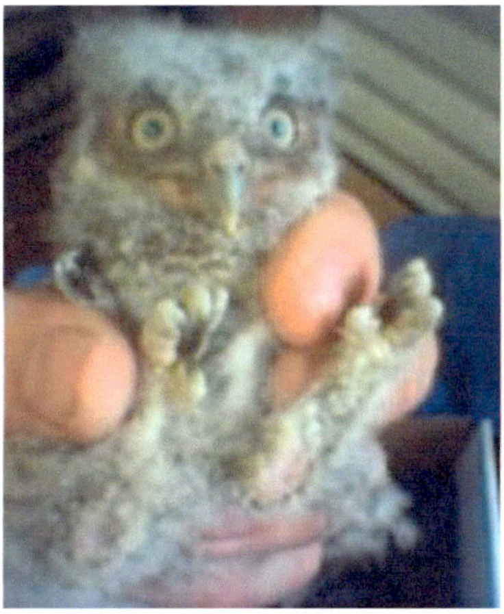

In another few weeks, the screech owl is full grown and showing its juvenile stripes. Its ears are fully tufted, and its claws fully functional. It is ready to fly!

Gray Phase Screech Owl Baby Fully Feathered.

Humans Raising Screech Owls

The most common problem with a human raising baby birds is imprinting on humans. All animals 'imprint." Imprinting is when an animal sees its parent and realizes that it is that kind of animal. If a bird imprints on its actual parents, then everything is good. If a bird imprints on a human, it can be disastrous. Imprinting makes the bird become accustomed to the presence of people and makes it assume it is a human as well. The dangers of imprinting are the following:

-The bird will not fear people, and therefore not stay away from civilization.

-The bird may attack people in an attempt to set itself as the dominant in that area.

-The bird will not identify with other birds of prey and may never mate or may even attack other birds.

-The bird may not hunt. It will expect food to be delivered to it.

To avoid imprinting, the baby bird must not see people. Use camouflage of some kind if you must be around the bird (for feeding, etc.) Especially do not let them see your face during feeding. Hand feeding should be done with either a puppet (looks like mommy) or with tools like large tweezers. If you save the bodies of rehab birds that do not survive, they can be made into 'puppets.' Basically, remove the innards and bones, dry the skin and form the bird into a wearable glove. This body can then be used to feed the infants with you camouflaged and attached to it.

It is okay to let them see you when you examine them, especially if you are restraining them. The idea that humans mean captivity and people poke and hurt birds will help make a bird ready for release.

The second most common problem is incorrect diet. In the wild, a bird of prey gets complete animals. The mother bird provides specific parts for the baby to eat. These parts generally are not just meat. Bone is one of the most important items in the baby bird's diet. Within 1 week a bird can become totally and permanently disabled by rickets. Rickets is a fatal calcium deficiency in the bones that lead to a slow and painful death.

Signs of rickets are:
-undersize bird
-bird that tends to lay down more than stand
-brittle bones; bird seems to be extremely fragile

Providing the bird with the correct diet prevents calcium deficiency problems.

-Use adult mice (or rats or chicks) cut into pieces for feeding; not pinkies (baby mice.)

-Dip all pieces of mouse into powdered bone. Bone meal is available in garden centers as calcium supplements.

If fed correctly and not imprinted, a baby bird can be fostered in with an adult of the same species and taught to hunt, and be released.

3- Flying With The Screech Owls

Screech owls ready to launch and fly!

Screech owls are able to fly quickly and silently throughout the woods and out in fields. They average 5 wing flaps per second, making their wings barely visible. Like all owls, they have special feathers that are fringed. The fringes are designed to silence their flight. Unless these feathers are wet, a healthy owl can fly in total silence. These feathers are unique to owls, and are very important because many owls hunt primarily by sound. When they are wet, the feathers do not muffle sound, so the flapping noise actually interrupts their ability to hunt.

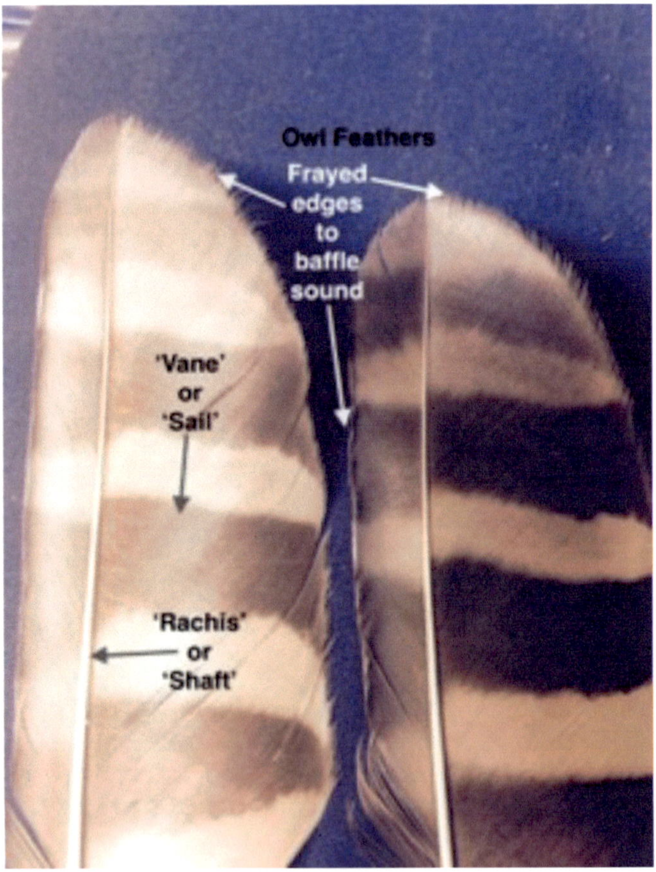

Owl Feathers Showing The Edges

Screech owls, like other birds, have unique wing bone structures to allow them to have enough strength to hold themselves aloft. Unlike mammals, their wrists and finger structures are compound bones for extra strength and flexibility.

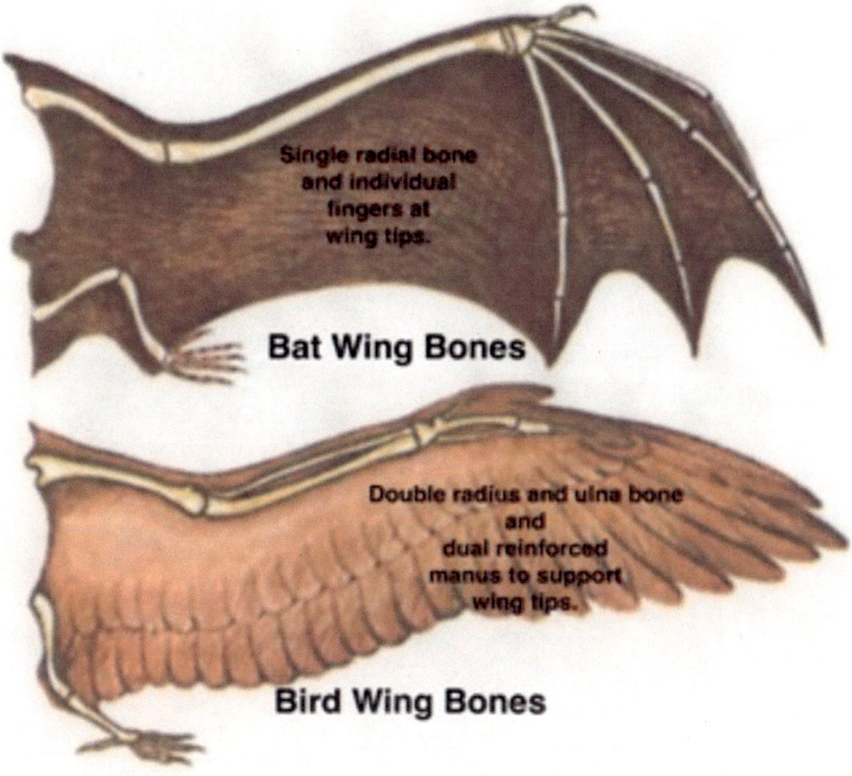

The entire skeletal structure of an owl is unique. Unlike other birds, they have a much longer neck, unique and powerful legs and feet, and a skull that is different from most birds.

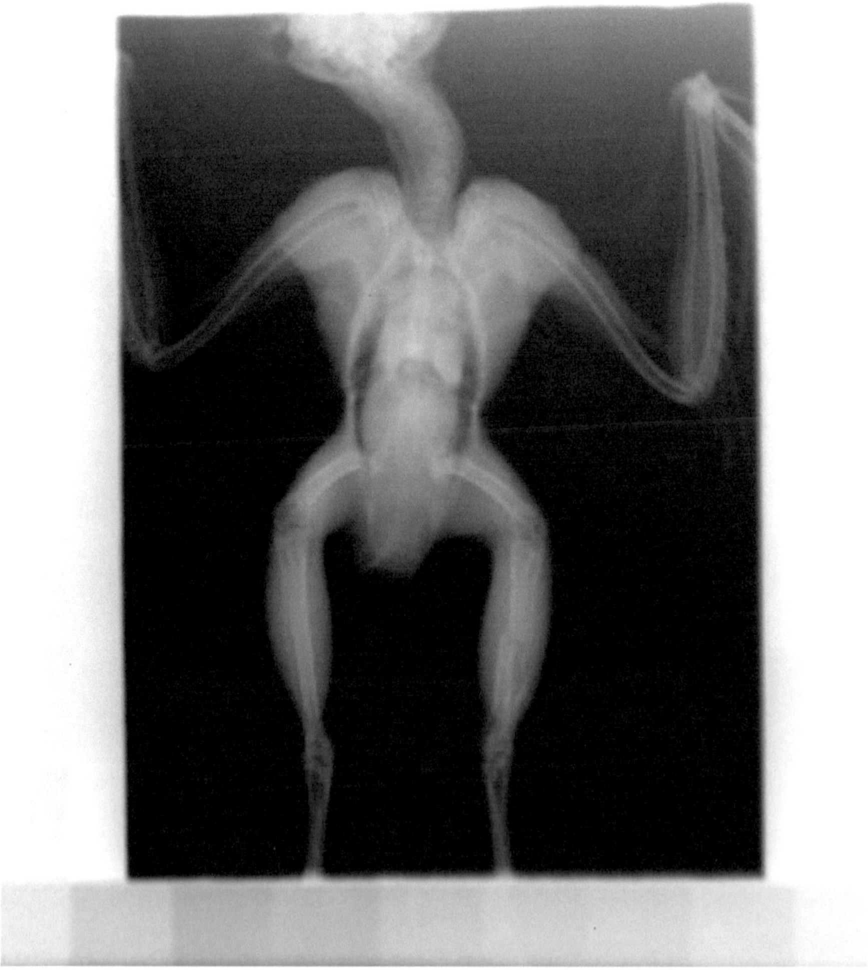

Owls Skeleton X-ray

They also all have their eyes mounted in unmoving bone tubes (sclerotic rings) giving them a more flat face and human appearance than other raptors.The feathers hide the shape of the skull and further the appearance of the facial disk, making it appear flat and human-like.

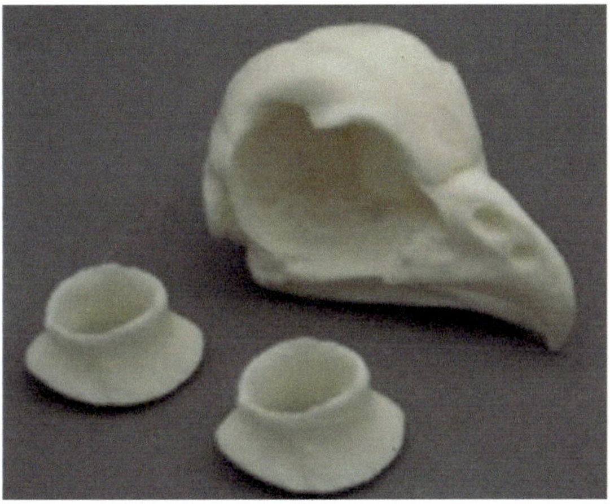

Owl skull model with model 'bone tubes' that support eyes.

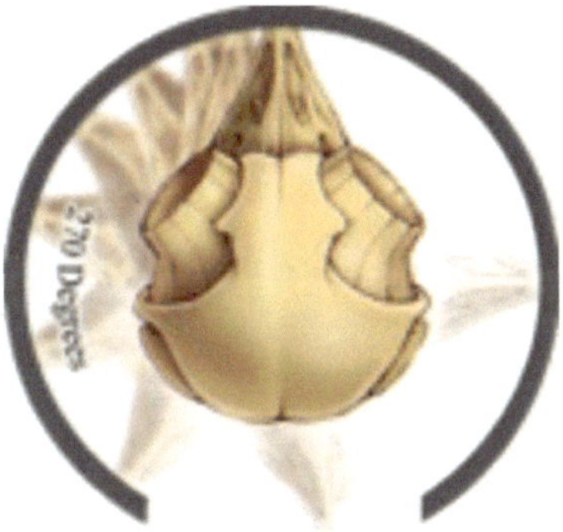

The extra long neck allows the skull 270 degree rotation.

Additionally, all owls have many physiological differences from all other raptors. First they have stomachs, not crops and gizzards. Their casts (pellets) are not just hair or feathers like those of other raptors, the owl's pellets have bones in them as well.

Second, their outer toes are hinged to be able to grip forwards or backwards. The three front toes and one rear toe that all raptors have is an option to these birds. However, most owls perch with two (2) toes forward and two (2) toes behind. This actually increases their grip strength. Owls generally have about a third (1/3) more grip strength than their hawk counterparts (e.g. a Great Horned Owl's grip strength is nearly 2000 lbs; a Red Tail Hawk's grip strength is 1500 lbs).

There are 226 uniquely different species of owls in the world. The US has over 122 species of owls, 3 of which are screech owls, according to owlpages.com, the current public authority on owls. Rather than get into arguments about which owls are different species, we can divide owls into 3 types, based upon their general physiological assets.

Owl Type 1 - Nocturnal or Night Owls - Nocturnal owls are usually the black eyed owls, but there are exceptions. For example, the Long Eared Owl is nocturnal, but it has bright yellow eyes. Nocturnal owls have twice the rods and 1/2 (half) the cones in their eyes when compared to humans. That means they see at least two (2) times the detail, but only half (1/2) the color. Their color vision is in the blue spectrum and partially in the ultraviolet. The ultraviolet light vision means they see a bit more at night than we do. Urine and other biological fluids glow in the dark to their eyes, so their vision helps them track prey, like the microtus vole which leaves a urine trail everywhere it goes.

Nocturnal owls also have much greater hearing than most other animals. The soft hair-like feathers on their face and around their eyes is called the "ruff." The ruff brings sounds to the ear using the entire surface of the face, helping the bird amplify and localize its hearing. According to experiments, the Barn Owl can hear the heartbeat of a mouse at 100 feet.

All owls hear better than humans, and most hear better than dogs, but the nocturnal ones hear so well, they can literally hunt in total darkness. YouTube has a video from BBC showing a great Grey Owl hunting through three (3) feet of snow. That bird is hunting by sound alone.

Nocturnal owls usually have a variety of feather colorings, most of which have at least two (2) or three (3) colors on the same feather. Camouflage is simple shading or dappling, not complex coloration like the Twilight Owls.

NC Night Owls include the Barn Owl (Tyto alba) and the Barred Owl (Strix varia). Both of these black eyed owls are generally nocturnal only. They are seen during the day only when they have babies in the nest crying for food or they have been rehabbed and adopted a daytime pattern due to their rehabilitation schedules.

Owl Type 2 - Crepuscular or Twilight Owls - Twilight owls are generally owls with yellow eyes. The yellow iris helps their eye gather light in lower light conditions. They also have twice the rods and half (1/2) the cones as humans do. Twilight owls are more dependent upon light to hunt and their hearing only slightly better than a dog's. Their ruff is usually much smaller, covering less than half their face.

Twilight owls are often more colorful, with the color designed to help them hide against trees and rocks. There can be as many as six (6) colors on one single feather in crepuscular owls. These owls will come out often in late afternoon, hunt during twilight, dusk, and early night, and then roost. They roost until just before sunrise, then go out and hunt again until mid-morning, and then go back to roost. They sleep in the middle of the day and the middle of the night, and are active around sunrise and sunset.

NC Twilight Owls include the Great Horned Owl (Bubo virginianus) and the Eastern Screech Owl (Megascops asio). Both of these colorful and camouflaged birds have feather tuft "ears" and bright yellow eyes.

Owl Type 3 - Diurnal or Day Owls - There are very few daylight owls. Daylight owls include Northern Hawk Owls, and Burrowing Owls. These are birds which are active during the day, but not usually active at night. Generally these owls depend upon sight, not sound for hunting and self-protection. Their hearing is only slightly better than that of a human.

*Special Note - No owl can actually win against a cat that is outdoors. An owl could win in an attack on a kitten, but generally speaking, cats kill great horned, barred, barn and screech owls on a regular basis. Anything else is an urban myth.

But where do screech owls fit into the 3 types? Screech owls physiologically are crepuscular, or Twilight Owls. The do not, however, always follow the rules of that type of owl. Screech owls often stay hidden during the day and do not leave their hiding place until after sunset. Unlike other crepuscular owls, they shun the daylight as much as possible, and never make their noises until after dark.

Unlike nocturnal, or night owls, Screech owls do not have especially good hearing. Their most common hunting ground in modern times is around streetlights and open yards and parking lots. The artificial lighting brings out their favorite prey, the large Polyphemus Moth and the Luna Moth.

Screech owls are usually about 5 inches from head to feet. They can stretch up almost another inch, or hunch down an inch or so. In NC there are red (actually more rust) and gray screech owls. As they get older, they both turn closer to brown. The duller the red color, and the more brownish the gray, the older the bird.

An Eastern Screech Owl measures about 5 inches head to toe.

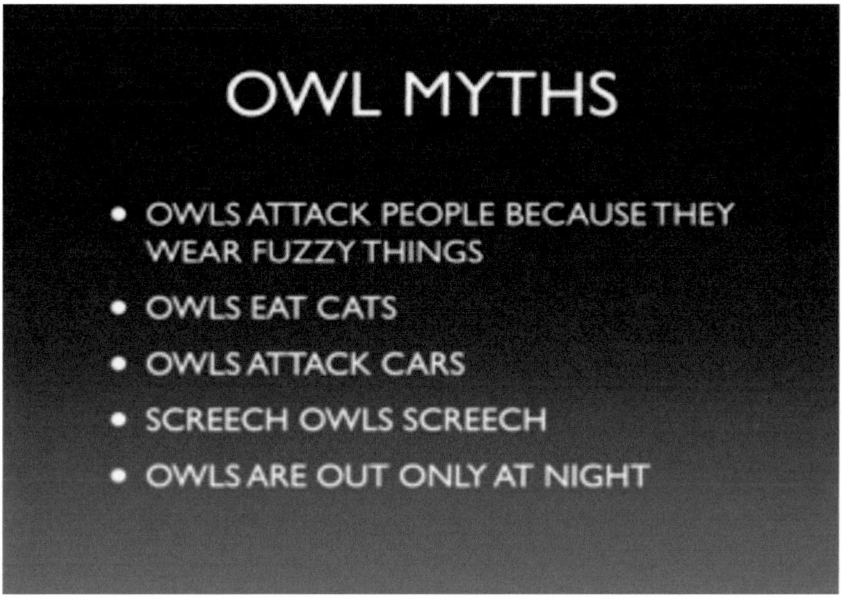

- Owls' main hunting sense is their hearing. While they can use their sight, they depend on sound to provide them with the main clues. Some owls can hear heartbeats of small animals; they are rarely fooled by fuzzy hair ties. The main reason owls occasionally attack a human is to protect their babies from a person near the nest.

- Owls do not hunt cats, but they will attack them to try to scare them away. Owls hunt mice for a living, cats do it for pleasure. When cats make it harder for owls to get mice, owls will dive bomb them and annoy them like a mockingbird or catbird. They are not hunting the cat; they are just trying to get it to leave.

- Owls' hearing is nearly parabolic. That means most of the sound they recognize comes from in front of them. They may not hear a car approaching from the side as they concentrate on the animal they are hunting across a busy street. When they slam into a car, or a car slams into them, it's a surprise to the owl.

- Screech Owls do not screech. The Eastern screech owls make a trilling noise and a horse whiney and sometimes an 'ooooh' sound, but they do not screech. The western screech owl makes a 'bouncing ball' sound, but again, no screech.

- Only about half the owls in the US are actually night owls. Many of them are twilight owls, and some of them are daytime birds. As a general rule, only the black eyed owls are night owls. The yellow eyed owls usually aren't.

Feeding Birds

Screech Owl - *Otus asio/Megascops Asio* -
(MALE = 140-210 grams, FEMALE= 150-235 grams)
Eats mostly rodents and insects, but will also take small fish, birds, and reptiles. In rehab, feed it 1-3 medium size (5 to 10 gram) mice daily or 1 day old chick (approx 30 grams) until it is ready for a flight cage.

According to **The Carolina Raptor Center** in Charlotte NC, Screech Owls should be fed between 18 and 27 grams of food each day.

CRC recommends skipping feeding one day each week unless birds are heavily molting, on the nest / have babies, or it is cold (below 45f). They also recommend adding vitamin supplements to the foods daily.

Birds of prey all have unique diets. Most species of raptors have a specific diet and must be fed the same items when in the hospital for rehabilitation. The prefabricated diets that can be purchased for the birds are not sufficient for long term use; they are for emergency use only. In almost every case, the birds must be fed the entire animal. Most birds will eat animals that are already killed. In reality, an injured bird is safer not having to chase the prey inside its rehab box. Fresh killed is the best for the animal, but if that is not available, frozen and thawed are often sufficient. You must remember, freezing and thawing will destroy some of the nutrients in the animal. Be sure to replace missing vitamins with vitamin supplements.

In very extreme cases, birds cannot digest the entire animal. The animals in these cases are often emaciated and in need of fluids and nutritional therapy. For maximum nutrition, the prey should be cut open and the organs removed. The weakened patient can get sufficient nutrition from the organs with less digestive stress. After the patient is stabilized, regular feeding can begin.

Birds of prey can often go for up to a week without eating. A healthy bird may be allowed to do this, but an injured bird must not. Force-feeding and tube feeding a predator is an experience that you cannot learn in a book.

It is extremely important that you watch an experienced rehabber force feed a bird before you do it yourself. One of the most difficult and dangerous parts of rehab is maintaining the bird's nutritional and fluid needs. Not only does the bird fight you to keep you from restraining it, but it also does not want you to open its mouth.

It is important to remember that you can kill a bird with an eyedropper of water. Birds have very limited ability to block water from entering their lungs if it is poured into their mouth. One of the first and most important things you must learn about force-feeding and watering a bird is how to keep them from drowning or suffocating from the procedure. It is not easy, and it has to be done exactly or it may be fatal.

The second thing to remember is that birds have powerful muscles and hollow bones. They can literally break their own bones struggling to escape you. Be sure you have the bird held correctly and firmly.

The third thing you must know is that holding a bird still to feed it can suffocate it. Birds do not have a diaphragm like we do. The muscles that they use to breathe are not protected by a rib cage. Holding them tightly can compress the bird's lungs and cause it to asphyxiate.

Always remember that each bird is different, and they will often show differences in attitude towards food. Some birds need to be force fed, while others just need to be left alone. If force feeding is necessary, do not force feed more than 1/2 to 2/3 of the normal food quantity. Force feed approximately half, and leave the other half in the cage to give the bird a chance to start eating on its own.

Emergency foods for raptors:

After excessive bleeding
If the bird has lost excess blood, feed them beef heart pieces along with the normal foods. The beef heart has extra iron and helps restore the blood.

For weakened birds
If the bird is excessively weak, feed them the organs and innards of a rodent. The nutrients are comparable to the entire animal, but they are much easier to digest for a weakened animal.

Emergency liquid nutrition
If the bird is too weak for any food, you can make an emergency nutrition solution. Take 1 quart of warm (body temperature) water. Add 1 teaspoon of salt and 3 tablespoons of sugar. Stir until completely dissolved.

This recipe makes a .5% saline 4.77% glucose solution for ingestion. It is only good for about 2 days, and must be kept refrigerated.

Special Considerations in Diet

The most common diet for raptors is rodents. Feeding a bird an all rodent diet will generally keep the bird healthy, but there does need to be some variety. Rodents are low in some nutrients that birds need. Some supplements are necessary if the bird is to remain healthy.

One common nutritional supplement is day old chicks. Day old chicks are chickens that are still young enough to contain their yolk sack. The nutrition in the yolk sack is one of the most concentrated vitamin sources for birds that exist. The only thing lacking in the nutrition from day old chicks is calcium.

Dusting food with powdered bone meal can add calcium to food. The bone powder can be purchased from either gardening stores or pet shops. This supplement is especially necessary for growing infants and birds with broken bones.

Insects are sometimes used as supplements to the diet of smaller raptors. Screech Owls, Kestrels, and other small raptors regularly eat insects as a large part of their diet in the wild. Crickets and mealworms are ready sources for this type of supplements.

It must also be remembered that water intake is an important part of the nutrition of a bird. Most raptors depend on the fluids in their prey as their water. They will drink from the water bowls only if they do not get enough water from their prey. Be sure that the food is not dried or dehydrated, and that there is a ready supply of water at all times.

Appendix A - Screech Owl Species Names:

Type of screech owl:

Eastern	(Megascops asio)
Western	(Megascops kennicotti)
Whiskered	(Megascops trichopsis)
Bearded	(Megascops barbarus)
Tropical	(Megascops choliba)
Puerto Rican	(Megascops nupides)
Pacific	(Megascops cooperi)
Balsas	(Megascops seductus)
Bare shanked	(Megascops clarkii)
Tumbes	(Megascops pacificus)
Oaxaca	(Megascops lambi)
Guatemalan	(Megascops guatamalae)
Long-tufted	(Megascops sanctaecatarinae
Columbian	(Megascops columbianus)
Koepcke's	(Megascops koepckeae)

Cloud forest	(Megascops marshalli)
White throated	(Megascops albogularis)
Cinnamon	(Megascops Petersoni)
Vermiculated	(Megascops vermiculatus)
Black capped	(Megascops atricapillus)
Montane Forest	(Megascops hoyi)
Southern Tawny-Bellied	(Megascops usta)
Peruvian	(Megascops roboratus)
Rio Napo	(Megascops napensis)
Northern Tawny Bellied	(Megascops watsonli)
Rufescent	(Megascop singens)
Santa Marta	(Megascops gilisi
Roriama	(Megascops roraimae)
Choco’	(Megascops centralis)

About The Author

Steve Stone has been training and working with raptors since the 1990's. He was primary caretaker of Beanca the Steppe Eagle for the last 14 years of her 47-year life.

Steve has both Federal and State certifications for the rehabilitation of all migratory birds. He is involved with each and every injured bird that comes through the American Wildlife Refuge. Steve has worked with many species of raptors that few in North Carolina have ever seen. Since he took over he has rehabilitated over 1200 birds with an average survival rate near 70%

Steve is also licensed for educational possession of migratory birds. Steve brings his educational animals to thousands of children each year. He also teaches park rangers and wildlife agents about raptors. To date his 1000+ educational shows have been seen by approximately 1.5 million people.

Steve is a member of IWRC, NWRA, and IAATE.

Made in the USA
Monee, IL
12 November 2026

67388635R00021